OUR SOLAR SYSTEM

COMETS AND METEORS

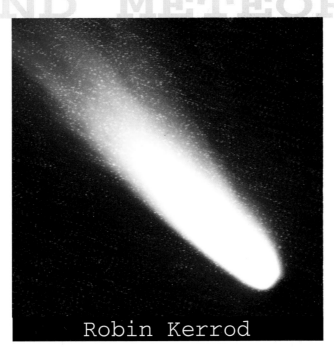

Robin Kerrod

Thameside Press

Distributed in the United States by
Smart Apple Media
1980 Lookout Drive
North Mankato, MN 56003

Text copyright © Robin Kerrod 2000

Managing editor: Veronica Ross

Designers: John Jamieson, Jamie Asher
Illustrator: David Atkinson
Consultant: Douglas Millard
Picture researcher: Diana Morris

Printed in Singapore

Library of Congress Cataloging-in-Publication Data

Kerrod, Robin.
    Comets and meteors / by Robin Kerrod.
        p. cm. -- (Our solar system)
    Includes index.
    Summary: Provides information about the origins, composition,
and appearances of comets, meteors, and asteroids.
    ISBN 1-929298-63-3
    1. Comets--Juvenile literature. 2. Meteors--Juvenile literature.
    [1. Comets. 2. Meteors.] I. Title.

QB721.5.K45 2000
523.6--dc21   00-024743

9 8 7 6 5 4 3 2 1

PHOTO CREDITS
Art Archive: 9b. Mary Evans Picture LIbrary: 16t. Nick James: 17.
Robin Kerrod/Spacecharts: 6, 18-19, 19t. MPIA/Spacecharts: 27t.
Martin Mobberley: 4-5. NASA/Spacecharts: 11t, 16-17, 20, 21, 26b.
Natural History Museum: 18c. NOAO/Spacecharts: 7t, 12-13, 15t.
Novosti/Spacecharts: 27b. Philippe Plailly/Eurelios/SPL: 26c.
RAS/Spacecharts: 7b, 17c, 22b. RGO/Spacecharts: 11b.
RO, Edinburgh/Spacecharts: 14b. Spacecharts: 8-9, 8b, 9t, 12t,
13b, 15c, 15b, 17t, 24, 25.

Some of the more unfamiliar words used in this book
are explained in the glossary on page 31.

CONTENTS

INTRODUCTION

Nearly every time you go stargazing, you will see little streaks of light in the sky. It looks as if some of the stars are shooting off to another part of the heavens, or falling toward the ground. That is why people call these streaks shooting stars or falling stars—but their real name is "meteors." Meteors are specks of matter from space burning up as they pass through the atmosphere.

These specks are some of the many lumps and bits of matter that are found in the space between the planets. Sometimes, bigger lumps pass through the night sky and are lit up by the Sun. We see them as comets, with a glowing head and a long tail fanning out behind. There are larger lumps still that we can't see, orbiting in the space between Mars and Jupiter. These are the asteroids.

NIGHT SKY SPECTACULARS

night sky spectaculars

Meteors and comets are among the most exciting sights in the night sky.

Usually, the changes that take place in the night sky occur gradually. The constellations, or star patterns, appear and disappear as the months go by. The planets wander slowly against the background of stars.

But meteors and comets appear and disappear suddenly. Meteors, can come and go in seconds—and they may be seen in any part of the sky. Exceptionally bright meteors may glow all the colors of the rainbow, and may even explode. We call them fireballs.

Catching comets

Comets take longer to come and go. These days, astronomers are not so surprised when comets appear because they have usually spotted them in telescopes months before. Most comets pass through the heavens quickly, and you can notice them changing positions among the stars from night to night.

Not all comets are bright enough to see with the naked eye. In fact, bright comets are quite rare. But stargazers in the late 1990s were fortunate to see very bright comets for two years running. They were Hyakutake in 1996, and Hale-Bopp in 1997. They both became brighter than all but the brightest stars and were truly spectacular.

◁ This spectacular comet of 1997, Hale-Bopp, hangs in the western sky just after sunset. It was one of the brightest comets of the twentieth century, as bright as the brightest stars.

Taking photographs

You can take photographs of meteors and bright comets with most cameras, but not on automatic mode. They must have a time setting. This will let you keep the camera shutter open for as long as you want.

You will also need two other pieces of equipment: a tripod and a cable shutter release. The tripod helps hold the camera steady, while the cable release stops you jogging it when you open and close the shutter. Use a moderately fast film, such as ISO 400.

To photograph a bright comet, point the camera at the comet and keep the shutter open for about a minute. If you keep it open for much longer, the stars in the picture will show up as little trails as they move.

Hit or miss

Photographing meteors is more hit and miss, because they can appear in any direction. The best approach is to point the camera to different parts of the sky in turn, keeping the shutter open for an hour or so each time. When the film is developed, you may be lucky to find meteor streaks on it—or you may not! Whatever happens, you will have some nice star trails.

◁ An astronomer photographing Comet Brooks in 1993 captured a fiery meteor trail.

STARS OF ILL OMEN

Most ancient people believed comets brought death and destruction.

People who lived long ago knew little scientifically about their world. And they knew even less about the heavenly bodies. However, they had a feeling that their lives were somehow ruled by what happened in the heavens. This belief became known as astrology.

When comets suddenly appeared without warning in the sky, many people often became scared. It seemed a very bad sign. This belief, that comets were bad news, lasted for thousands of years.

Chinese broom stars

As early as the 600s B.C., Chinese writers said that comets were "vile stars." They caused emperors to die, uprisings among the people, floods, drought, and diseases. The Chinese were very skilled observers of the heavens, and seemed particularly interested in comets. Their earliest records of comets date as far back as 1400 B.C.

The ancient Chinese called comets "broom stars," thinking that their tails looked like the bristles on a broom.

Other peoples have called them "stars with feathers" and "hair stars." This idea is echoed in the modern word comet, which comes from the Greek word *coma*, meaning hair.

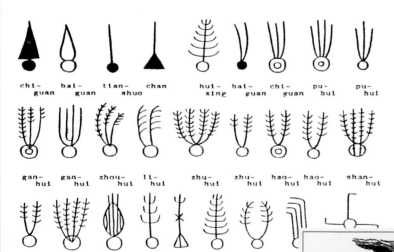

△ Ancient Chinese astronomers recorded many kinds of comets. These sketches are based on drawings dating back to 186 B.C.

▷ The brilliant 1664 comet, as seen at Nuremberg, Germany.

Nord

West

Ost

Sud

Im Jahr Christi, 1664. den ¼⁄24 Decemb: in der Nacht gegen Tag, nacht 5. der Kleinnern Uhr, ward in def H. Röm, Freyen Reichs Stadt Nürnberg, diefer Erfhröckliche Comet Stern wie hier Abgebildet Zuerfehen.

More bad news

Comets have been blamed for all kinds of disasters. Roman writers said they caused wars. In the Middle Ages, it was said that comets always appeared when kings and princes died.

This certainly proved true in 1066, when a comet appeared in the sky at the time of the Battle of Hastings. King Harold and the English army were fighting the invading Norman army led by William the Conqueror. The English lost the battle after Harold was killed. The 1066 comet later proved to be a regular visitor to Earth's skies, and is now known as Halley's Comet.

△ **The Bayeux Tapestry shows the comet of 1066 above King Harold on the throne of England.**

△ **Edmond Halley realized that some comets visit our skies regularly.**

Epidemics from space

Some astronomers have suggested that comets carry tiny, primitive organisms (life forms), such as bacteria and viruses. As they pass near to the Sun, the microorganisms drift into space. In time, they enter the Earth's atmosphere and cause worldwide epidemics, such as influenza.

The fall of the Aztecs

In Mexico, in 1519, Montezuma, the Aztec king, saw two brilliant comets. He took this as a sign that the white-bearded god Quetzalcoatl was coming. Not long after, the bearded Hernan Cortés arrived with his Spanish soldiers. Thinking that Cortés was the god, Montezuma welcomed him. But, within a short time, the Spaniards had all but wiped out the Aztec civilization.

HEADS AND TAILS

The tails of comets can sometimes reach 93 million miles in length.

A bright comet is a spectacular sight. Its bright head looks like a flaming arrow pointing towards the Sun. Its tail fanning out behind stretches for millions of miles across the starry sky.

But what exactly is a comet? Mostly, it is a great cloud of gas and dust, like the smog that forms in polluted cities on Earth. The cloud is lit up by sunlight and shows up brilliantly against the blackness of space.

Only deep inside the comet's head is there a solid bit, called the nucleus. Compared with the size of the whole comet, the nucleus is tiny. In most comets, it is only between about $\frac{1}{2}$–6 miles across. The bright 1997 comet Hale-Bopp was unusual in having an exceptionally big nucleus—as big as about 25 miles across.

Dirty snowballs

The nucleus of a comet is often described as a "dirty snowball," because it seems to be made up mainly of water, ice, and dust. When a comet enters the inner solar system, the Sun's heat makes the ice on the surface of the comet evaporate, or turn to vapor (gas). The gas spurts out of the nucleus in jets, carrying dust with it. Large comets like Hale-Bopp give off vast amounts of gas and dust—as much as 1,100 tons of dust and 165 tons of gas every second.

The gas and dust form a cloud around the nucleus, hundreds of thousands of miles across. This cloud is the glowing head of the comet, or coma. The coma is surrounded by an even bigger cloud, made up of hydrogen atoms.

gas jets

nucleus

△ Jets of gas shoot out of the tiny nucleus of a comet, forming a huge glowing cloud around it. Sunlight forces the gas into a long, broad tail.

▷ **Comet West was a striking sight in spring skies in March, 1976. Here we see its two tails clearly—the blue one being the gas, or ion tail.**

The hydrogen cloud is invisible in ordinary light, but shows up in pictures taken in ultraviolet light.

Telling tails

The most spectacular part of a comet is its tail— or rather, tails— because most comets have two tails.

How do comets grow tails? Jets of gas and dust spurt out of the comet nucleus on the hottest side, which faces the Sun. As they spurt out, they come up against the solar wind. This is a stream of particles flowing out from the Sun. The wind sweeps the gas and dust backward.

Soon, two distinct tails develop, one made by the gas, one by the dust. The particles in the solar wind hit the gas particles and make them electrically charged. They become what are called ions, and start to glow. They form the gas, or ion tail. Sunlight pushes the dust particles into another tail, called the dust tail.

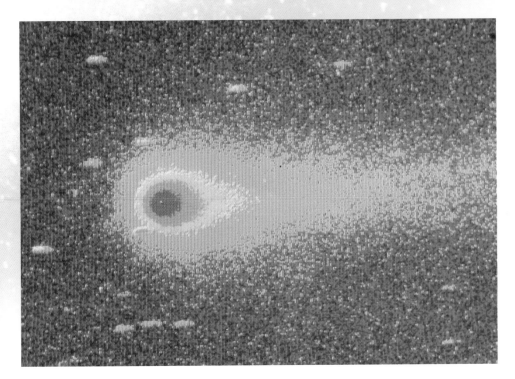

◁ **A false-color view of Halley's comet in 1986, shows different levels of brightness.**

IN FROM THE COLD

Comets are born in the deathly-cold outer reaches of the solar system.

Comets seem to appear suddenly in the heavens, and then brighten as they move toward the Sun. As they move away from the Sun, they fade and disappear from view. It is this sudden appearance and disappearance that frightened people in the past. But comets do not really suddenly appear and disappear. Like other bodies in the solar system, they follow orbits, or paths through space, as they travel around the Sun.

For most of the time, their orbits take comets away from the Sun into the dark depths of the solar system. Little heat or light reaches them there, and so they remain frozen solid and invisible.

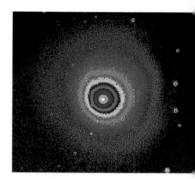

△ **When first spotted, a comet looks like a glowing ball.**

It is only when comets start to close in on the Sun that they begin to light up. They also start to melt and release their cloud of gas and dust, and grow tails. With most comets, this happens when they approach the orbit of Mars, 155 million miles away from the Sun. After they loop around the Sun and travel back into space, they disappear at about the same distance.

tails grow shorter

Sun

comet orbit

tails grow longer

△ **Comet tails always point away from the Sun as the solar wind pushes them. Comets develop the longest tails when they get closest to the Sun.**

Curving orbits

Comets have different kinds of orbits from the planets. The planets travel around the Sun in nearly circular orbits. They also travel in much the same plane (flat sheet) in space, and in the same direction. And from the Earth, we see the planets move through the same part of the sky.

Comets could not be more different. They have all kinds of oval and curving orbits. They can appear and disappear anywhere in the sky and move in any direction.

▽ **This false-color picture shows Halley's comet in 1910.**

The long and the short

Some comets travel in orbits that bring them back near the Sun in a short time, or period. One, called Encke's Comet, returns every 3.3 years. Astronomers call it a periodical comet, and term it P/Encke. Halley's Comet, P/Halley, returns to Earth's skies after every 76 years or so. In general, comets with return periods up to 200 years are called short-period comets.

Most bright comets that appear have not been recorded before and have long periods. They have wide, curving orbits that may not bring them back near the Sun for thousands of years. The 1997 comet Hale-Bopp will probably not return for 4,000 years—the 1974 comet Kohoutek, won't be back for 75,000 years.

17:56 UT

◁ **The space probe SOHO spots two comets diving into the Sun. This happens when comets get too close.**

EDMOND HALLEY'S COMET

Halley's Comet returns to Earth's skies about every 76 years.

In January, 1681 a brilliant comet blazed in Earth's skies. Among the many who watched it was an English astronomer named Edmond Halley. After another bright comet appeared the following year, he worked out its orbit. Then he checked its orbit with the orbits of other comets that had appeared in the past.

Halley found that comets that had appeared in 1607 and 1531 seemed to have similar orbits to the 1682 comet. He was convinced that these comets were one and the same, returning to Earth's skies about every 76 years. In 1705, he published a book about comets and predicted that the comet of 1682 would return in 1758.

Halley died in 1742. Sixteen years later, on Christmas night 1758, a German astronomer found the comet Halley had predicted. Since then, Edmond Halley's Comet has returned to Earth's skies three times—the latest being in 1986. It will return next in 2061.

Target Halley

When Halley's Comet returned in 1986, it proved to be disappointing for most people. It was only just visible to the naked eye in certain parts of the world. But astronomers studied it closely through their telescopes. And space scientists launched probes to spy on the comet from close quarters.

The most successful probe was *Giotto*, launched by the European Space Agency (ESA), which went closest to the comet. Russia sent two probes (*Vega 1* and *2*), and so did Japan (*Sakigake* and *Suisei*).

◁ **Halley's Comet, photographed through a telescope from Australia in March, 1986.**

Halley in close-up

After an eight-month journey across the solar system, *Giotto* homed in on Halley's Comet in March, 1986. It flew to within 370 miles of the comet, taking spectacular pictures. They showed the nucleus and spotted fiery jets of gas and dust streaming from it.

The nucleus measured about 9 miles long and 5 miles across. It twisted and turned slowly as it traveled. The surface was surprisingly dark and rough. *Giotto* images showed what looked like hills and craters.

△ A broad tail fans out from the head in April, 1986. The comet is now heading away from the Sun.

Halley's nucleus proved to be made up mainly of water ice, along with mineral dust, and carbon compounds. It was probably a coating of sooty carbon material that made the surface look so dark. Water vapor was also the main substance in the gas cloud surrounding the nucleus. Carbon monoxide and carbon dioxide were also found, along with formaldehyde. The discovery of formaldehyde was very exciting. It is an organic compound that is also found on Earth. It may be that comets carry organic compounds around the solar system.

△ The European space probe *Giotto*, which passed only about 370 miles from Halley's comet.

3443

◁ *Giotto* spots glowing gas jets spurting out of the comet's dark nucleus.

CATCH A FALLING STAR

Specks of space dust burn-up in a flash when they enter the atmosphere.

The space between the planets is full of bits of matter. They may be as small as specks of dust or as large as boulders. Astronomers call these bits of matter "meteoroids."

Where do meteoroids begin? Some are the little bits left over from the time the solar system was born, about 4,600 million years ago. Others are the bits comets leave behind as they travel near the Sun.

The Earth passes through swarms of meteoroids all the time as it travels through space. And those nearby are attracted by the Earth's pull, or gravity.

△ **An artist recorded a brilliant fireball in the skies of outer London, England in February, 1850.**

They plunge towards the ground, some at speeds of 60,000 mph or more. As soon as they hit the gases in the Earth's atmosphere, the meteoroids start to heat up because of friction. They get hotter and hotter and start to burn up. As they burn, they leave behind fiery trails. It is these trails that we see in the night sky as meteors—popularly called shooting or falling stars.

Meteoroids big and small

Most of the meteoroid particles that cause meteors are about the size of a grain of sand. They burn up by the time they have reached a height of about 50 miles above the Earth, leaving nothing but fine ash behind.

△ **Not all streaks in the sky are made by meteors. The spacecraft _Apollo 8_ made this one when it returned to Earth after flying round the Moon.**

ing

▷ Bright meteors flash through the sky in November, 1998 during the Leonid meteor shower.

▽ A large meteor blazes a trail through the sky and heats up so quickly that it explodes. Some bits may survive and fall to the ground.

Light in the zodiac
The dusty particles that fill the space between the planets can cause faint glows in the night sky. This can happen just after the sky darkens in the evening or in the dark sky just before sunrise in the morning. We call these glows "the zodiacal light," because they can be seen in the constellations of the zodiac.

But most particles are smaller, even microscopic in size. They do not burn up as meteors but drift slowly through the atmosphere and eventually settle on the ground.

Showers of meteors
On most clear nights, you should be able to see up to about 10 meteors every hour. They can come from any direction and are known as "sporadic meteors."

But, at certain times of the year, many more meteors appear than usual—sometimes hundreds an hour. When this happens, we call it a meteor shower.

A shower happens when the Earth travels through a stream of particles left behind by a comet. In a shower, the meteors come from the same part of the sky. The shower is named after the star constellation where the meteors appear to begin. For example, in August every year, a shower comes from the direction of Perseus, and the meteors are called the Perseids.

STONES FROM THE SKY

Bits of rock and metal from space continually bombard the Earth.

Some of the meteoroids captured by the Earth's gravity are much bigger than average. They only partly burn up as they travel through the atmosphere. The bits that are left fall to the ground as meteorites.

Thousands of meteorites have been found throughout the world. Most of them are little bigger than pebbles. Only a few really big ones are known. The biggest is one found in Namibia, in South-west Africa. It is called the Hoba meteorite and it measures more than 8 feet 2 inches across and weighs about 66 tons.

Large meteorites hit the ground so hard that they dig out pits, or craters. More than 120 craters are known on the Earth, the largest being Meteor Crater in the Arizona Desert.

▷ **Arizona's famous Meteor Crater is the finest on Earth. It measures 4,149 feet across and 574 feet deep. It may be up to 50,000 years old.**

Irons and stones

The Hoba meteorite is made up of metal, like all the large meteorites that have been found. Metal meteorites are known as irons, because they are made up of iron, with some nickel. The two metals form an special alloy. When a piece is polished and treated with acid, a fascinating criss-cross pattern shows up.

◁ **An iron meteorite found near Meteor Crater. When polished, it shows a typical criss-cross pattern.**

Most meteorites are made up of minerals, like the rocks and stones on the Earth. That is why they are called stones. The most common types are known as chondrites. They usually contain rounded lumps of silicate minerals (chondrules). Some are rich in carbon compounds.

A much rarer group of meteorites contain both metal and mineral fragments. They seem to be a combination of the other two types, and are called stony-irons.

Meteorites young and old

Scientists have methods of finding out the age of meteorites. Most meteorites turn out to be older than any rocks found on the Earth's surface. They are about 4,600 million years old. This means that they date back to the time when the solar system was born.

▷ **This meteorite was found buried in ice in Antarctica in 1981. Scientists think it probably came from Mars.**

A few meteorites are much younger. Some found in the Antarctic are only about 1,300 million years old. And scientists now reckon that these came originally from the planet Mars. They were probably flung into space after a huge asteroid hit the planet. Other asteroids may have come from the Moon, as they have a similar make-up to Moon rocks brought back by the Apollo astronauts.

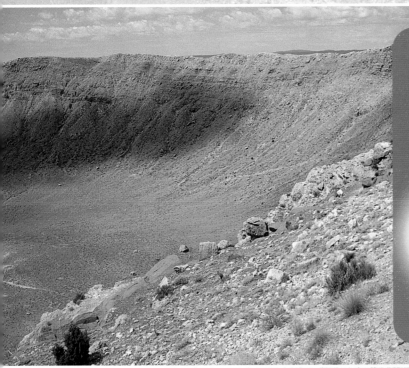

Gaining weight

The Earth gains up to about 88,000 tons in weight every year from the meteorite material that rains down on it from outer space. Most of this material is in the form of tiny particles of space dust and ash (from burning meteors). These particles are known as micrometeorites.

SHAPING OTHER WORLDS

Meteorites have bombarded and shaped many other bodies in the solar system.

On the Earth today, we find only a few craters made by meteorites. When the Earth was young, swarms of rocky lumps smashed into our planet. But, over the years, the craters they made have disappeared. They have been worn away by the action of the weather or covered up by upheavals in the Earth's crust.

The rocky swarms also bombarded the other bodies in the solar system. And most of them still bear the scars of this bombardment.

△ Saturn's Moon Enceladus has only a few craters because its surface is so young.

◁ Our own Moon is covered with thousands of craters. Its surface hasn't changed for billions of years.

A battered face

Meteorites have largely shaped our companion in space, the Moon. Thousands of craters, large and small, cover the surface. The largest ones, such as Bailly and Clavius, are more than 155 miles across. They have high walls, which descend in terraces to deep floors. Often, there are mountain peaks in the middle of the craters.

Even the huge plains we call seas came about because of meteorite bombardment. Particularly large rocks slammed into the Moon, creating mountainous walls. The force of the impacts caused volcanoes to erupt. The lava they gave off flooded the huge craters, creating the vast flat seas.

Assault on the planets

We can also see the effect meteorites have had on the other rocky planets—Mercury, Venus, and Mars. Mercury is covered all over with craters. They have not worn away over the years because the planet has no atmosphere. And without an atmosphere, it has no weather to wear away the surface. Also, there have been no volcanoes or ground movements to help cover up the old craters.

The next planet out, Venus, has no really old craters, and very few new ones. The reason is that the planet is still very active, like the Earth. Volcanoes are found all over Venus. In recent times, they have erupted time and time again, flooding the landscape with lava. Any old craters have been covered up. Venus also has a thick atmosphere, which helps protect it from small falling meteorites. They burn up before they reach the ground.

Meteorite craters are also found on Mars, but mainly in certain areas. Other areas are flat plains regions, which have been created by erupting volcanoes and also possibly by ancient oceans.

▽ **Phobos, Mars's largest Moon, is only about 12 miles across, but it is peppered with craters.**

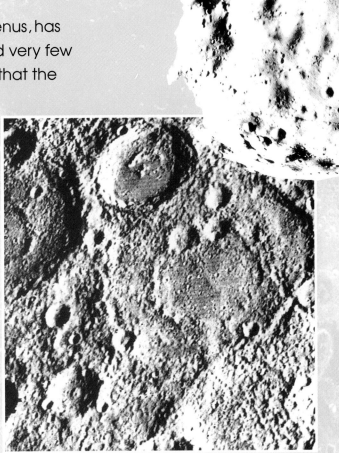

◁ **Mercury is the most heavily-cratered planet in the solar system. Few areas have escaped bombardment by meteorites.**

The giant planets

Meteorites have had no visible effects on the giant planets beyond Mars—Jupiter, Saturn, Uranus, and Neptune. This is because these planets have no solid surface and are made up mainly of gas and liquid. But most of the solid moons of these planets are heavily cratered.

MANY MINI-PLANETS

A swarm of rocky bodies circles the Sun between Mars and Jupiter.

The four rocky planets—Mercury, Venus, Earth, and Mars—lie close together in the center of the solar system. There is then a huge gap of nearly 370 million miles before the next planet, Jupiter.

Astronomers first realized there was a gap when they began calculating planet orbits around 1600. The German astronomer Johannes Kepler suggested that there ought to be another planet in the gap—but 200 years went by and no planet was found.

Then, on the first day of January, 1801 an astronomer from Sicily named Giuseppe Piazzi spotted one. He called it Ceres after the patron goddess of Sicily. It looked like a star and so could not be very big. Soon three other "mini-planets" were discovered in the gap: Pallas (in 1802), Juno (1804), and Vesta (1807).

asteroid belt

Jupiter

asteroids

△ **Asteroids show up as streaks in this photograph.**

Asteroid swarms

Since then, thousands of "mini-planets" have been discovered. Astronomers call them minor planets or asteroids. Most of the asteroids circle the Sun in a broad band between the orbits of Mars and Jupiter called the asteroid belt. The middle of the belt lies about 250 million miles from the Sun.

Ceres

Vesta

Pallas

△ **Ceres is nearly twice as big as the next largest asteroids, Vesta and Pallas.**

Sun

Mars

◁ The asteroid belt lies between Mars and Jupiter, and is about 125 million miles wide. It contains at least 40,000 asteroids.

Not all asteroids stay in the asteroid belt. Some wander as far as the orbit of Saturn, while others wander in toward the Sun. One, called Icarus, gets closer to the Sun than the planet Mercury. Several asteroids wander near the Earth's orbit. They are known as near-Earth objects or NEOs.

The celestial police

In September, 1800 a Hungarian nobleman named Baron Franz von Zach decided to organize an international search party for the "missing" planet between Mars and Jupiter. They became known as the celestial police. One of the "police," Heinrich Olbers, discovered the second and fourth asteroids—Pallas and Vesta. Another searcher, Karl Harding, discovered the third asteroid—Juno.

The centaur asteroids

As well as the ordinary asteroids, there are other bodies known as the Centaur asteroids. They have been found circling the Sun as far out as the orbit of Neptune. U.S. astronomer Charles Kowal found the first one, Chiron, in 1977. Astronomers have also found similar objects even further away —in a remote region called the Kuiper belt.

Sizing the asteroids

The first asteroid discovered, Ceres, is by far the largest. It measures nearly 620 miles across, making it less than a third the size of the Moon. The next largest asteroids (Pallas and Vesta) are only about half the size of Ceres. In all, there are only 25 asteroids bigger than 125 miles across.

LOOKING AT ASTEROIDS

Most asteroids are shapeless lumps, battered by repeated collisions.

Most asteroids are much bigger than comets. But they never come close enough to the Earth to be visible to the naked eye—except for Vesta. This asteroid sometimes becomes bright enough to be visible if you know exactly where to look for it in the sky. It shows up as a pinpoint of light, like a distant star.

Studying asteroids

Powerful telescopes can detect thousands of asteroids down to a few miles across. They give out only a feeble light,

but astronomers can tell a lot from it. From tiny changes in an asteroid's brightness, they can tell how fast it is spinning and can estimate the asteroid's shape. Most asteroids appear shapeless. Only the biggest, Ceres, appears to be round like the Moon. Astronomers also get information from the spectrum of an asteroid's light. This is the rainbow-like band of color obtained by passing the light through a prism or similar device. From the spectrum, astronomers can work out what the asteroid contains.

△ **Many small craters cover asteroid 951, named Gaspra. About 11 miles long, it was photographed by the *Galileo* space probe in 1991.**

Different types

Asteroids are made up of three main kinds of substances—rock, carbon, and metal. Astronomers group them into three types —S, C, and M. Most asteroids in the inner part of the asteroid belt are S (silicaceous) types, made up of silicate rock. They tend to be reddish in color. Most asteroids in the outer part of the belt are C (carbonaceous) types, containing large amounts of carbon and carbon compounds. They are dark in color, and some are as black as soot.

> ▷ This photo of Vesta was taken by the Hubble Space Telescope. The different colors show that the surface is made up of different materials.

5 cm

2 in.

△ Scientists think that this meteorite, found in Western Australia, came from Vesta. It seems to have the same chemical make-up.

Asteroids galore

A special telescope called LINEAR, operated by the U.S. Air Force in New Mexico, has spotted an average of more than 1,000 new asteroids every month since it went into operation in March, 1998! It has also recorded several hundred NEOs. Originally, LINEAR was set up to track satellites and man-made debris in space.

The M (metallic) types of asteroids are not so common. They are made up mainly of the metals iron and nickel. Astronomers think that they are probably the centers, or cores, of larger asteroids that were broken apart in collisions.

Asteroid images

No one knew what asteroids looked like until 1991. In that year, the space probe *Galileo* sent back pictures of Gaspra. The probe snapped Ida two years later. Both were shapeless lumps of rock, pitted with craters. Since then, the Hubble Space Telescope has sent back images of Vesta and other asteroids. Some asteroids have also been imaged from the Earth by using radar.

◁ Asteroid 243, named Ida, is nearly 35 miles long. Amazingly, it even has a moon, Dactyl, just under 1 mile across.

COSMIC COLLISIONS

Lumps of rock wandering in space can cause disasters.

In July, 1994 astronomers saw something no one had seen before—the bombardment of another world from space. More than 20 pieces of a comet named Shoemaker-Levy 9 crashed into Jupiter, one by one. As they smashed into the planet, great fireballs were created, which astronomers could see in their telescopes.

This event set people wondering:. "Could something like this happen on the Earth?" "The answer is that yes, it could." And in the past it has —with terrible consequences.

△ Scientists carefully excavating dinosaur bones. All dinosaurs died out about 65 million years ago.

Death to dinosaurs

About 65 million years ago, over two-thirds of all the plants and animals that lived on the Earth at the time became extinct, dying out. Among them were the "terrible lizards," the dinosaurs. Most scientists now think that this mass extinction happened when the Earth was hit by an asteroid. They reckon that it might have been as big as 9 miles across. And they believe that it struck the Earth in what is now the Yucatan Peninsula in Mexico.

△ The fragments of Comet Shoemaker-Levy 9, pictured in May 1994. In two months, they will be smashing into Jupiter.

◁ **Three massive hot spots show where fragments of Comet Shoemaker-Levy 9 have hit Jupiter in July, 1994.**

Near misses

In the history of the Earth, 65 million years ago is quite recent. There is still a chance of asteroids, or comets, hitting the Earth again. In 1908, an asteroid or small comet exploded in the atmosphere in Siberia, northern Asia. The blast damaged over 1,170 square miles of trees in the forest near the Tunguska River. If the explosion had taken place over Moscow, most of the city would have been destroyed.

The Earth-grazers

The asteroids that are the biggest threat to the Earth are the NEOs, or near-Earth objects. There are hundreds of such bodies, and more are being found all the time. In May, 1993 an NEO came within 93,000 miles of our planet—a very near miss indeed.

Shock waves

Here is what might have happened. The force of the asteroid smashing into the ground dug out a vast crater, hundreds of miles across. Millions of tons of rock were smashed to pieces and blasted into the atmosphere. Shock waves rippled round the Earth. Tidal waves hundreds of yards high sped across the oceans.

So much fine dust was created that it hung in the atmosphere for months. It blotted out the Sun and plunged the Earth into darkness. Most plants died, because plants need sunlight to make their food. And without plants to feed on, most animals died too—especially big ones like the dinosaurs. They needed vast amounts of food to stay alive.

▷ **The forest region in Siberia, Russia, was damaged by an exploding asteroid or comet in 1908.**

NOTHING BUT THE FACTS

METEOR SHOWERS

Shower	Constellation	Begins	Ends	Max. number per hour
Quadrantids	Boötes	Jan 1	Jan 6	110
Eta Aquarids	Aquarius	May 2	May 7	20
Delta Aquarids	Aquarius	July 15	Aug 15	35
Perseids	Perseus	July 25	Aug 18	65
Orionids	Orion	Oct 16	Oct 26	30
Leonids	Leo	Nov 15	Nov 19	varies
Geminids	Gemini	Dec 7	Dec 15	55

THE BIGGEST METEORITE CRATERS

Name	Place	When discovered	Diameter (feet)
Meteor Crater	Arizona	1891	4,149
Wolf Creek	Australia	1947	2,788
Henbury	Australia	1931	656
Boxhole	Australia	1937	574
Odessa	Texas	1921	557

THE BIGGEST METEORITES

Name	Type	Weight (tons)	Where found
Hoba West	Iron	66	Near Grootfontein, Southwest Africa
Ahnighito	Iron	33.5	Cape York, West Greenland
Bacuberito	Iron	30	Mexico
Mbosi	Iron	28.6	Tanzania, East Africa
Agpalic	Iron	22	Cape York, West Greenland

WHAT A SHOWER

The number of meteors seen in Leonid meteor showers varies widely. At the most, hundreds can be seen every minute. Maximum showers tend to occur about every 33 years—and there were brilliant—showers in 1966 and 1999.

METEORITES ON ICE

One of the best places to find meteorites is on the far southern continent of Antarctica. Scientists have already recovered more than 10,000 specimens from there. One of the richest regions is near the Allan Hills in Victoria Land.

MOON METEORITE

On January 18, 1982 the U.S. geologist John Schutt found a meteorite, ALHA81005, in the Allan Hills region. When they examined it closely, it proved to be full of tiny glass spheres, just like rocks brought back from the Moon. This meteorite probably came from the Moon.

Comet notes

MANY HAPPY RETURNS

Encke's Comet circles the Sun every 3.3 years—the shortest period of any comet. It was named after the German astronomer Johann Encke, who calculated its orbit and predicted its return in 1822. By 2000, it had been seen on its last 58 returns. It easily beats Halley's Comet, which has been seen on only 28 returns (since 240 B.C.).

MESSIER'S TABLE

One of the great comet hunters of the 1700s was the French astronomer Charles Messier. Like most astronomers of the day, he was constantly confusing comets (which often look rather like fuzzy patches in the sky) with star clusters and nebulae (which also look like fuzzy patches). So he decided to make a list of nebulae and clusters and note where in the heavens they were found. Then, he could eliminate them when he was observing. Astronomers still often identify a star cluster or nebula by the number Messier gave it in his list. For example, the Orion Nebula, is referred to as M42.

RECORD BREAKER

Messier was a prolific discover of comets, with 13 to his credit. But the record is held by the French astronomer Jean-Louis Pons. By the time he died, in 1831, he had discovered no fewer than 36 comets.

BREAKNECK SPEED

The brilliant comet of 1965, Ikeya-Seki, passed less than 310,000 miles from the Sun. As it approached, the Sun's powerful gravity accelerated the comet until it was traveling at a speed of more than 0.93 million mph. Like most comets that travel so close to the Sun, Ikeya-Seki was broken up by gravitational forces into two or three fragments.

Asteroid notes

LIKE A STAR

In 1802, the German astronomer Heinrich Olbers, a member of the "celestial police" who were hunting for new planets, turned his telescope on the newly discovered body Ceres, and found another new body in the same part of the sky. He called it Pallas. Astronomers agreed these objects were not true planets. The English astronomer William Herschel suggested that they be called asteroids, a word meaning "star-like."

EMPTY GAPS

In a number of places in the asteroid belt, there are gaps where no asteroids are to be found. They are called Kirkwood gaps, after the U.S. astronomer Daniel Kirkwood who predicted their existence in 1857. They occur because of gravitational forces set up by the giant planet Jupiter.

GRAZING THE SUN

The orbit of the asteroid Icarus takes it within 18,600,000 miles of the Sun, within the orbit of the planet Mercury. At close approach, the temperature of its surface must be more than 900°F. It is named after a character in Greek mythology who escaped from captivity by means of wings held together by wax. But he flew so close to the Sun that the wax melted, and he plunged to his death.

TIME LINE

65,000,000 B.C.
A massive asteroid smashes into the Earth, creating dust clouds that blot out the Sun for months. Hundreds of species of plants and animals die out.

50,000 B.C.
A huge meteorite hits the Arizona desert about this time, digging the Arizona Meteor Crater.

240 B.C.
The comet we now call Halley's Comet is recorded for the first time.

1066
A comet appears at the Battle of Hastings, another sighting of Halley's Comet.

1682
The English astronomer Edmond Halley suggests that the comet of this year is a regular visitor to Earth's skies and predicts that it will return in 1758.

1758
As Halley predicted, the comet returns and is henceforth named after him.

1801
The Italian astronomer Giuseppe Piazzi discovers the first asteroid, Ceres, on January 1.

1861
The U.S. astronomer Daniel Kirkwood suggests a relationship between meteors and comets.

1908
A comet or asteroid explodes near the Tunguska River, Siberia, laying waste to vast areas of forest.

1950
The Dutch astronomer Jan Oort proposes that a great "cloud" of comets lies at the edge of the Solar System. The U.S. astronomer Fred Whipple suggests that a comet's nucleus is like a "dirty snowball."

1966
The Leonid meteor shower on November 17 is spectacular, with thousands of meteors an hour visible.

1977
The US astronomer Charles Kowal discovers a far-flung asteroid, named Chiron.

1986
Halley's Comet is seen best in the Southern Hemisphere. The Giotto probe flies near it (372 miles) on March 14.

1991
The *Galileo* probe takes the first picture of an asteroid, Gaspra.

1993
In May, an asteroid passes within 93,000 miles of the Earth—a near miss.

1994
The many pieces of Comet Shoemaker-Levy 9 crash into Jupiter in July.

1996
Comet Hyakutake makes a spectacular visit to Earth's skies. NASA scientists suggest that a meteorite from Mars (ALH84001) contains microscopic fossils. Others disagree.

1997
Comet Hale-Bopp blazes in spring skies, becoming one of the brightest comets of the century.

1999
Space probe *Deep Space 1* swoops within 9.3 miles of an asteroid named Braille in July. The Leonid meteor shower puts on a spectacular show in some parts of the world on November 17–18.

2000
The NEAR probe goes into orbit around Eros on February 14, beginning a year-long study of the tiny asteroid.

GLOSSARY

asteroids
Small lumps of rock and metal that circle the Sun in a band—the asteroid belt—between the orbits of Mars and Jupiter.

atmosphere
The layer of gases around the Earth or another planet.

axis
An imaginary line through the middle of a spinning body, around which the body spins.

coma
The glowing head of a comet.

comet
A small lump of ice and dust that starts to shine when it gets near the Sun.

constellation
A group of bright stars appearing in the same direction in the sky.

crater
A circular pit in the surface of a planet or moon. Most craters are made by meteorites, but some are the mouths of ancient volcanoes.

dust tail
One of the two tails of a comet, made up of dust, which reflects sunlight.

fireball
An exceptionally bright meteor, which may explode.

gas tail
One of the two tails of a comet, made up of glowing gas.

gravity
The force of attraction that every body has because of its mass.

head
The front part of a comet, properly called the coma.

interplanetary
Between the planets.

irons
Meteorites made up mainly of iron.

meteor
The fiery trail left by a meteoroid burning up in the atmosphere.

meteorite
The remains of a large meteoroid that hits a planet or a moon.

meteoroid
A piece of rock or metal found in space.

meteor shower
A time when more meteors than usual can be seen.

minor planets
Another name for the asteroids.

moon
The common name for a satellite.

Near-Earth Objects (NEOs)
Asteroids that pass close to the Earth.

nucleus
The small solid part of a comet, within the head.

orbit
The path in space one body follows when it circles another.

period
A length of time in a regularly repeating cycle, such as the period of a comet (the time it takes to circle the Sun).

planet
One of the nine bodies in space that circle the Sun.

probe
A spacecraft sent to explore other heavenly bodies, such as planets.

satellite
A small body that orbits around a larger one; a moon. Also the usual term for an artificial satellite, a spacecraft that orbits the Earth.

shooting star
A popular name for a meteor.

solar
To do with the Sun.

solar system
The Sun and the family of bodies that circle around it, including planets, comets, and asteroids.

solar wind
A stream of particles that the Sun gives off.

stones
Meteorites made up mainly of rock.

tail
The part of a comet that streams away from the head.

universe
Space and everything that is in it—galaxies, stars, planets, and energy.

INDEX